SECRETS TO
UN-HACKABLE
FACEBOOK ACCOUNT

SPECIAL TRICKS TO SECURE YOUR ACCOUNT FROM LOCAL & WORLD CLASS HACKERS

Goodness Kosi

Although the author and publisher of this book have made every effort to ensure that the information in this book was correct at press time, the author and publisher do not assume and hereby disclaim any liability to any party for any loss, damage or disruption caused by errors or omissions, whether such errors or omissions result from negligence, accident or any other cause.

Copyright © 2019 Goodness Kosi

All rights reserved. No part of this book may be reproduced, stored in a retrievable system or transmitted in any form or by any means, without the prior written permission of the publisher, except in the case of brief quotations embodied in critical articles or reviews.

"One problem with technology is that, our solution becomes deadly when it enters the wrong hands"
_Goodness Kosi

Table of contents

1. First thing to do
2. Keep your password safe!
3. Do not help a Facebook hacker
4. Disconnect third-party services
5. Browsing With a different device
6. Do Not Allow your Curiosity to Mislead You
7. Keep Your Contact Info Up-to-date
8. Take Advantage of the inbuilt Facebook security system
9. How to Enable "Two-factor Authentication"
10. How to set up extra Security
11. How to choose 3 to 5 trusted friends who can help you recover your account
12. Facebook is doing their part, do yours to secure your account
13. What to do if your account has already been hacked

Foreword

One Tuesday evening, I logged into my Facebook account, but discovered a strange message from one of my friend.

I knew she wouldn't send such a message to me. And my mind flashed to the fact that her account has been compromised.

Luckily, I helped her to secure and regain control of her account. But, the hacker has already sent out messages on her behalf. And her only option was to send an apology message to the affected friends.

On another occasion, I discovered a strange name from the list of my active chats. I couldn't remember when I accepted such a friend request. So, I decided to check the profile and saw only few old posts and two unrecognized images.

I was very curious and decided to say "hi". I was shocked when I opened the message box and saw lots of chats which have taken place between me and the unrecognized profile.

I was like: how come? As I checked the messages, I discovered that the account belongs to my friend, but has been hacked and most of his pictures and posts have been removed. So, I had to call him immediately and we recovered the account.

These scenarios are just few. There are other worst cases of hacks and hundreds of hacks occur each day.

Probably, you've heard of more incidences of Facebook account hacks or you have been a victim either once or more.

It's a painful ordeal especially when they turn your account to a mess. They may do a lot of silly things with your account. They may even defraud your friends and make them think you did.

If you are using your Facebook account for business or to promote a brand, this is more reason you need to read this book from cover to cover.

Because they are hunting influencer's accounts, they want to get hold of your followers and defraud them in your name.

They can even sell your pages to your competitors.

Well, I compiled this book for your own benefit.

But, it can only help you if you don't skip any part of this book because there are secrets you need to learn and apply so that you can outsmart any kind of Facebook hacker.

Whether they are local, international or world class hacker, none of them will be able to hack your account if you pay attention to the details of this book and *apply* what you learn.

But, is it practical to say "no hacker can crack your account if you apply everything you learn in this book"?

Yes of course!

The reason remains that Facebook is the largest social media in the world.

Guess what?

They have a lot of enemies, but Facebook is always stronger than their enemies.

Do you know that Facebook servers are nearly impossible to hack?

I am not saying they cannot be hacked but, they are nearly impossible to hack.

The reasons are because they have the world best cyber security experts, they spend a lot of money on server security and maintenance. They also

SECRETS TO UN-HACKABLE FACEBOOK ACCOUNT

have hackers and security researchers around the world who reports any security vulnerabilities or weaknesses and get monetary reward.

In short, Facebook know their enemies and will never allow their enemies to bring them down.

So, if Facebook are doing their part very well, why are there still successful Facebook account hacks?

Let's dive in and find out why and how to end it.

Introduction

You know? About a decade ago, Facebook was majorly a tool for fun but now more than that.

Facebook has improved so much and it is now a very big business tool.

In fact, it means different things for different faces, because beauty is always in the eyes of the beholder.

For some, it is a playground. For others, it is a business environment, a money making machine and a learning center.

If that's the case, your account is worth being protected from anyone who wants to breach your privacy or take over your position on Facebook.

This book contains all the steps and strategies to make sure your account is **SAFE** and **UN-HACK-ABLE.**

This book is practical, so be ready to practice what you learn.

Chapter 1

First thing to do

Make sure you are using a strong password!

Basically, Facebook will allow you to login by combining one of the 3 username options.

That is to say: you can log into your Facebook account by using one of the following combinations:

A. Your Username and Password
B. Your Phone number and Password
C. Your Facebook's E-mail and Password

Now, for some people with low retentive memory like me, (**Smiles, ** *No, I don't have low retentive memory*)

Maybe, you don't want to stress your brain. You don't have enough time to cram long passwords. And you hate the long process associated with retrieving a forgotten password.

Therefore, you choose an easy password so as to eliminate all the stress.

That's a bad approach but, if you are one of the hundreds of millions of Facebook users who use a weak password, don't worry.

Just keep reading, you'll discover not just how to create a strong password, but also how to make your Facebook account un-hack-able.

But, before we do, let's find out possible paths through which your account can be compromised.

So, how do hackers operate?

Well, they know that most human beings are lazy and always prefer easy paths to harder paths.

Therefore, most people will likely choose passwords that they can easily recall.

Records also show that this is true.

Back in the days when Facebook stores a hashed version of their users password in a database

Most local hackers or novice hackers hold on this fact and apply a manual trial and error method.

They may combine your Username with your Phone, Email or Username as the password.

In short, your account can easily be hacked by anyone, if your password is your email, your phone number, birthday or Username!

Now, let's look at this example:

A Facebook user with the username Johnny Mark registered with the phone number +2348222222.

After registration, Facebook assigned him *"facebook.com/johnnymark_6"* as a unique ID and *johnnymark_6@facebook.com* as his Facebook email.

Since Facebook allows you to log into your account using your email, phone number or username.

It is obvious that, Johnny Mark can log into his account using any of the following:

SECRETS TO UN-HACKABLE FACEBOOK ACCOUNT

Jonny Mark + password

+2348222222 + password

johnnymark_6@facebook.com + password

Now, assuming Johnny Mark chooses his email to be his password, do you notice how anyone can login and take over his account without stress?

Look, if you know the dangers of your account being hacked, your password will never be anything that can easily be guessed by anyone.

Your birth date, your friend's name, your nickname, name of a place or anything that can be found in a dictionary is not a good choice for password.

If you check online, you'll see some suggestions like;

Change the vowels in the name which you want to use as a password to its numeral equivalent.

That is, Johnny Mark will change to J4hnnyM1rk. Or, United States will change to Un3t2dst1t2s

Looks great right?

That's a very bad approach too.

Since choosing a strong password is the first step to make your Facebook account un-hack-able, let's dive into the process and create our first strong password.

Note: Strong passwords are: the combination of the following;

SECRETS TO UN-HACKABLE FACEBOOK ACCOUNT

1. Uppercase letters (A - Z)
2. Lowercase letters (a - z)
3. Numerals (0 – 9)
4. Special characters (, . ! @ $ % ^ & * ( ) _ - + = / \ | " : ; ' < { > ~ `) All the symbols in the closed braces are examples of special characters.

To create a very strong password, we have to choose some characters from Capital letters, Small letters, Numerals and Special characters. Choose anything, from each of the categories and mix them together in any order.

This is an example of what you may get:

ACbz12;?
Mix it in any order and you will get something like this **z?A1bC;2**

Did you see any difference in the position of the characters?

They are no longer the same thing and none of them is predictable.

Awesome: Right?

You can follow the same process and make it up to 20 digits, 15digits. 8 digits are ok too, but 6 digits are not that ok even if Facebook allowed you to proceed with a 6digit password.

Now, we've learnt how to create a very strong password. Play with it. Create your own password.

Warning: Do not use the password I created in our example. Only follow the process and create yours.

Chapter 2

Keep your password safe!

Like I said earlier, choosing a strong password is the FIRST step to an **UN-HACKABLE** Facebook account.

And the last article pointed out that Strong passwords are the combination of uppercase letters, Lowercase letters, Numerals and Special characters in any order.

But the fact is that, if you don't keep your password safe, it is useless no matter how strong it is.

Certainly, no one will likely go to his timeline to tell his friends about his password, except you were social engineered to do so!

How?

Some group of hackers may come to a group and say something like:

"Wow!
Do you know that, if you type your password here, Facebook will mask it?
Look at mine;
********"

When you check the comments section, you'll see confirmation comments from their members.

One of them may write a plain number.

After sometime, he'll write ******* and say, Wow! It works only if you type your real password!

He may also add *"Facebook algorithm is great!"*

Hey my friend, please do not type your password there.

Facebook does not mask any password.

In fact, if you try it, you may not have a chance to delete your comment because the hackers will immediately log into your account and lock you out.

What about using your exact Facebook password to login other websites?

Did you get the question? Please read again.

Using your Facebook password as the password for other website, is it advisable?

No, not at all

If you have been doing this, you may need to change your Facebook password as soon as possible!

Avoid using your Facebook password to register an account in other websites.

Obviously, hacking is of different grades. And some senior hackers do more than you may imagine.

Here is what they do:

SECRETS TO UN-HACKABLE FACEBOOK ACCOUNT

You know?

There are numerous websites on the internet and most of these websites store your password in their database as is or after hashing it: although, this is considered a bad practice, but most developers used to get it wrong. (Facebook was one of those websites that stores your password by hashing it, but not anymore.)

Ps: A database is like a memory where your information is stored on the internet. But, due to the increase in cybercrime, it is considered a very bad practice to store passwords in the database whether it is hashed or not.

Well, some hackers take the bad practice of some web developers as an advantage.

They will first of all hack in to the website which stores passwords in their database.

If successfully hacked, they will get hold of the passwords with other relevant data. Next, they will reverse the passwords if they were hashed and use it to get into your account if you are the type that uses the same password everywhere in the internet.

Other hackers will follow an easier approach.

They will create a membership website with a hidden aim of collecting passwords.

If you are using the same password over and over again, just imagine the risk!

If you are the type of person that uses the same password everywhere on the internet, the risk of your accounts being compromised is extremely high.

SECRETS TO UN-HACKABLE FACEBOOK ACCOUNT

So, what are the key points?

=====

1. Create a strong password

2. Do not share your Facebook password with anyone and do not use it to register in any other website.

3. When you change your password, do not use it again next time.

Keep in mind that hackers are smart, and if you need to prevent them, you must be smarter!

Chapter 3

Do not help a Facebook hacker

Avoid unknowingly helping a hacker compromise your account

Obviously, cannot come up with a reasonable solution if we don't understand what causes the problem.

This same ideology applies to what causes Facebook accounts to get hacked.

A hacker is in a mission which you are actively against. But you may unknowingly be in support and also be the commander who is giving them a go ahead order.

How?

Let's first understand the problem so that we will be able to fish out the loopholes.

One of the popular methods used by hackers is called "phishing" and this is the king and the easiest method which they use to get into your account and surprisingly, they complete this action with your permission.

That is to say: you just helped them but within you, you thought you were helping yourself. Sounds funny?

I guess. But it is a serious issue.

So, how do they do this?

They'll simply send you a link, possibly a video clip, an image, a document or whatever.

Most times, they package the information in such a way that you become very curious like a monkey that heard a strange sound.

SECRETS TO UN-HACKABLE FACEBOOK ACCOUNT

It is bait in a hook!

For instance: take a look at these sample links:

1. https://fakebook.com/easy-way-to-make-money-online

2. https://faecbook.com/amazing-nude-girl

3. https://facebook.com.profileboaster.com

Be sincere, did you notice the problems with the above links?

1st link was actually: faKebook.com, check again

2nd link was actually: fAEcbook.com, check again

3rd link fools one to believe the contents are from Facebook

No, rather facebook.com as appeared in the 3rd link is a sub domain of profileboaster.com. This means that when you click in the link, it will take you to a page in the website profileboaster.com

Note: In every URL, after "https:// or http://", the last name + dot (com or au or ng or org, etc.) which appears before the first forward slash "/"is the name of the website.

The reality is that when you click on such link, the page that prompts out will look exactly as the Facebook login page but as soon as you type in your password and click submit, you are deliberately giving your login details to some hoodlums in the other room.

Take a time and look at the address bar(the top part of your browser containing the URL), you'll notice that the resultant page which looks

SECRETS TO UN-HACKABLE FACEBOOK ACCOUNT

exactly as a Facebook login page is actually coming from an address different from that of Facebook.

The URL in your address bar may appear like the one in the above screenshot; *http://facebook.com.awoofsecretinfo.com*

Run away and do not enter your login details.

In the above URL, the website name is awoofSecretInfo.com with a sub domain name facebook.com so as to confuse a victim. So, a victim types in their password hoping that they are logging in to Facebook, not knowing that the page is fake and that he is consequently giving out his login details to a fraudster.

The fraudster will take the details, log in to your account and lock you out.

To avoid this, you should always be at alert whenever you are asked to enter your login details.

The image below shows a real Facebook login page. Look closely at the address bar and compare with the fake page above.

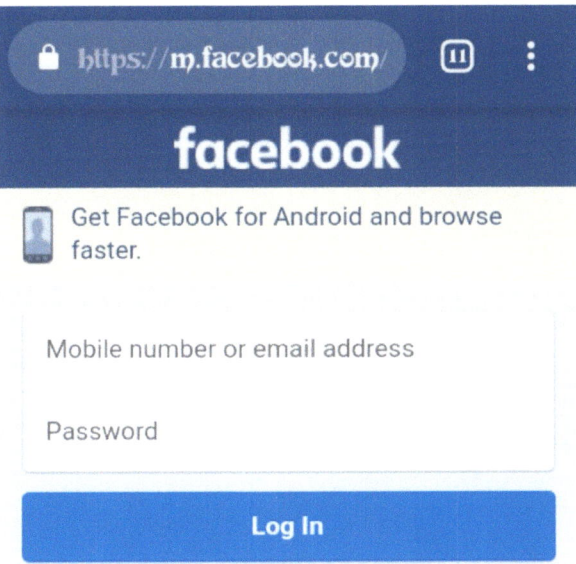

In the image above, can you see the forward slash which follows immediately after "*.com*"?

The forward slash which appeared immediately after *https://m.facebook.com* is the reason I concluded that the above screenshot is a real Facebook login page.

It's obvious that on mobile devices, the address bar is always short and shows only 20 to 23 first characters of a web address. Therefore, one can easily be fooled to believe that he is browsing the real Facebook login page.

Now, compare the image below with the previous images. Look at URL and observe the **hyphen** which appeared immediately after the **.com**

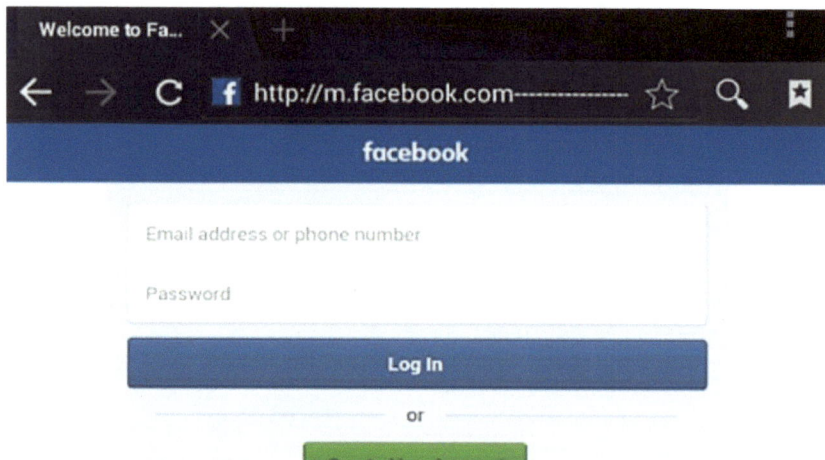

The above technique is called **URL-Padding** and it is a hacking technique used by world class hackers.

When you click on a link which they sent you, you'll see a page like the one above asking you to login.

You may not get suspicious because your mobile device will be showing you a minimum viewable URL of 20-23 characters.

After the hyphens, you'll now see the real address through which your login details will be forwarded to.

This remains the reason I will always say: be in your senses before clicking on a link.

SECRETS TO UN-HACKABLE FACEBOOK ACCOUNT

If you are browsing your Facebook account and suddenly, a link you clicked on prompts you to login again,

Examine the page or leave it entirely. It may be a plot to hack you.

Do you know the worst case?

These days, hackers may not be able to hack your Facebook account but can achieve it by simply hacking into your device.

Your device can be hacked once you click on a link.

I just hope you got it right.

I said, your device can be hacked immediately after you click on a link.

Just on a click, your device can be controlled by someone else.

They can track your keystrokes. The hacker can view your browser history and perform so many other functions like turning on your webcam.

Just know that when you click on a suspicious link, you are leaving a passage for a hacker to achieve his aim.

Assuming the link you clicked on, downloaded a key logger on your device, it means that the hacker will be able to see everything you typed in your device. Both pages you visited, ATM card details you typed in, plus other sensitive information.

Some key loggers are also installed in USB devices, so be careful.

SECRETS TO UN-HACKABLE FACEBOOK ACCOUNT

Here's what you can do

=====

1. Always be in your senses before clicking on clips, links or images you don't know their source.

2. Examine the link before clicking on it. If you can, check the link on Google search first.

3. Do not click on any suspicious link!

4. When you click on a link, image or video clip and it requires you to enter your password and username, please, please and please: do not give in if you don't want your account to get hacked.

5. Always use an updated version of your browser or Facebook app

6. Avoid app installation from unknown or not trusted publishers.

7. Do not associate every USB device that you see with the device you use to log in to your Facebook account.

8. Avoid visits to not trusted websites.

9. At best, install the best antivirus programs you can find.

Chapter 4

Disconnect third-party services

If you are not familiar with the term "third-party services" as it relates to Facebook, you may be wondering what they are and how to disconnect them.

Third party services are all the services that are not directly provided to you by Facebook and its company, but Facebook allowed other companies to provide those services to you.

So, any application which is not developed by Facebook, but functions on Facebook is a third-party app.

Any time you connect your Facebook to a third-party app or services, it accesses your profile or other information from your account.

This presents security risk.

The information gathered by the app may be used to hack your account. In minor cases, it can expose your photos and details to potentially harmful people.

For instance, **Quizzstar** is a quiz app that works on Facebook.

The app requires your permission to gather your profile information.

If the permissions are not changed through your Facebook account settings page, the app will continue to gather all the profile information you granted it access to.

That's exactly how all third-party apps work.

Generally speaking, it is not essentially bad to connect your Facebook account to a third-party app or services. But it is worth knowing that each one you connect to leaves a backdoor to your social Facebook account.

SECRETS TO UN-HACKABLE FACEBOOK ACCOUNT

To control the backdoors, it is essential to run a regular audit of all the services you connected and removing the ones which you no longer use or which are no longer updated.

The reason is that even if the apps are genuine, their database may fall prey to someone who isn't. So, the less exposed you are, the better.

If you ask me for an advice, I will tell you to avoid any application that claims to store your passwords for you.

Avoid it totally please.

Do you know that some of those apps are owned by professional hackers?

Even the genuine ones are not very safe.

The reason is because Facebook security is very tight, and the only way a hacker can penetrate into your account is by grabbing your password!

He can always have access to your username, phone number, Facebook unique ID and he may be able to guess your security answers and get them right except that he cannot get your password if you know how to keep them strong and safe!

Now, they know these. They know they can grab your password by breaking into the database of password storage apps because you are too busy to cram your password.

With these facts, don't you think that it is better to tell Facebook that you forgot your password and change it to a new one whenever you want to log in than to store your password in a password storage app?

Even Facebook no longer store your passwords in any of their database or server.

What?

How then do they recognize you entered the right password?

Well, they process it in such a way that no one, even their own engineers will know your password except you shared it with someone.

If you are a programmer, you may already have an idea of what I'm talking about.

But, you could be doing it in a wrong way.

In fact, I can prove it to you. But if you would like to know the exact process that Facebook experts use to process the users password without having to store it anywhere. Hook up with me after you read this book, although I would have loved to explain everything but, it's out of scope of this book.

So, let's go back to the major thing.

How to remove third-party services

1. Log in to your Facebook

2. Go to Settings

3. Under the section "Security", click on "Apps and Websites"

4. Click on "Logged in with Facebook", when it opens, you will see a list of active, expired and removed third-party services associated with your account.

5. Go ahead, select and remove the third-party services you no longer need.

SECRETS TO UN-HACKABLE FACEBOOK ACCOUNT 28

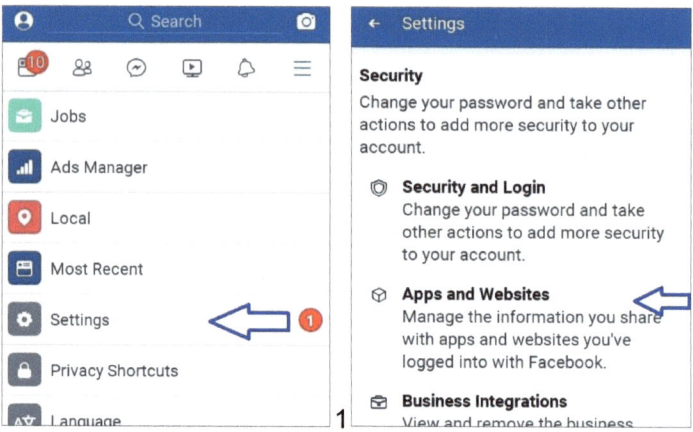

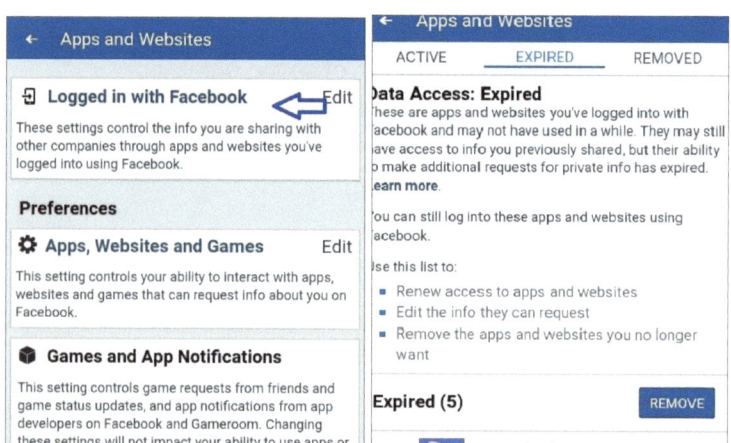

Note: After removing the services, the websites or apps will no longer gather your private information, but may still have info you previously shared with them.

Chapter 5

Browsing With a different device

Have you ever browsed your Facebook account using a device which is different from yours?

Probably, your own device is faulty and you logged into your Facebook account using your friend's handset or through a computer in a public cyber café.

What do you do when you are done?

You logged out right?

Well, some people will log out, and others may forget to log out. But logging out is not the real deal. This is because; some devices will automatically save your login details without your permission.

If this happens, your account can be compromised after you take your leave.

So, next time you log in to your account using someone else device:

Do these to keep your account safe.

=====

1. It is possible that the device may have a key logger installed. If it is a computer, use an on-screen keyboard when typing sensitive information. If it is a mobile device, do not proceed if you suspect a key logger is installed.

2. If the device asks for a permission to save your login details, click "No/ Decline".

3. Once you are done, logout and clear the "cookies".

4. When you later login to Facebook using your own device, go to Settings > Security > Security and Login >

   Here, you will see all the devices through which your Facebook account is being accessed. It shows you the name of the device, its version and location.

   That's amazing!

   So, if someone has been browsing your account without your permission, you will see it there too.

   You may decide to "Logout of all sessions" or "logout that particular device which you previously used to access your Facebook". You will also need to logout all unrecognized devices.

   When you do this, Facebook will require your password whenever someone wants to log in to your account through those devices.

   But, there is a little glitch here.

   What if your password was automatically saved in those devices?

   You have to clear the cookies in those devices if they are accessible to you, if not; the best thing to do is to change your password.

Besides, you may be wondering what cookies are. Simply say: cookies are bits of information about: you or your actions in a particular website; which are saved by the web server in your browser. No, I'm not talking of biscuits. (**smiles**)

SECRETS TO UN-HACKABLE FACEBOOK ACCOUNT

To clear the cookies, follow the steps in the bellow images:

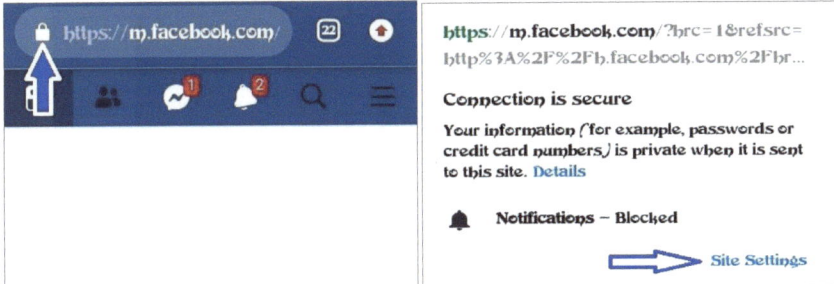

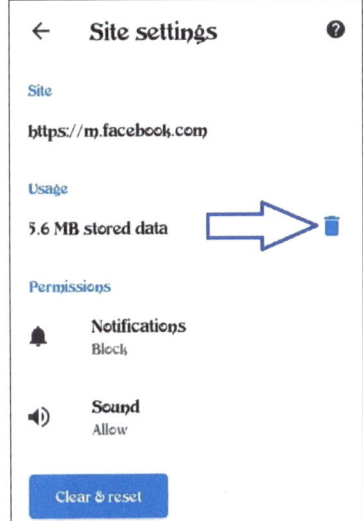

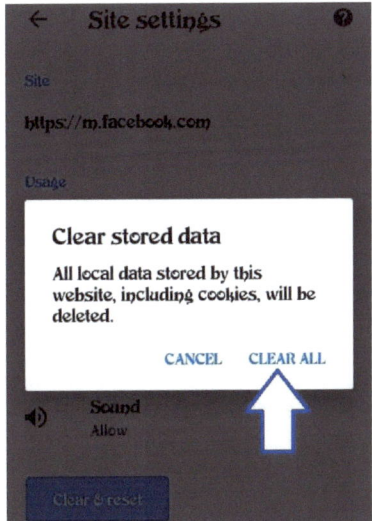

Chapter 6

Do Not Allow your Curiosity to Mislead You

Some of the successful Facebook account hacks are what the owner of the account brought to them self, probably due to their curiosity or jealousy.

Maybe, you are in a relationship.

You have a partner whom you care so much about. In fact, you always wish to know what he or she is doing.

You want to know who's chatting her and why. You'll always be like: "who's that guy that usually leaves darling comments on your wall?"

Probably, no answer would satisfy your curiosity.

And now, you don't want to continue giving out the expression which reflects your fears. So, you hit Google for help.

With fast fingers, you typed "*How to spy on my friend's Facebook messages without their permission*".

Opps! You are bringing doom to yourself.

Probably, some parents are also victims.

They want to spy on their children by any possible means. Actually, they genuinely want their children to browse safe and hangout with the right online friends.

So, they visit Google with such words like: "*How to spy on my child's social media activities*".

Surely, there are millions of results for anything you search on Google and majority of the resultant pages for the above search will claim to provide you with a smart solutions.

But, are they for real?

Sadly, not all of them are genuine solutions.

Some unethical hackers are online too, posing as genuine helpers. When you find them,

Here is what happens:

They will promise you to help you spy on your friend or child Facebook activities without them ever finding out. And you'll be like: "*Wow, this is what I'm looking for!*"

They may also promise to give you an easy to use Facebook hacking tool.

All are lies!

Instead of helping you, they are either going to take you through a very long survey which will make you open a lot of pages filled with advertisements.

Or

They will require you to login with your own Facebook login details. And if you try this, you are the first dupe.

Next, they will ask you to invite 2 or 3 friends whom you trust so much that they won't get suspicious.

Laughs

These friends are the inline dupes

Did you just notice how the curiosity and jealousy of one person become a threat to his own Facebook account and also a proposed threat to that of his friends?

What you can do:

=====

1. Always tame your curiosity and jealousy.
   Is spying on someone more important to you than securing your account?

2. When you click on a link and you are asked to provide your Facebook login details, stop and rethink.

Chapter 7

Keep Your Contact Info Up-to-date

When I was a child, I don't like to dispose any of my old belongings despite the fact that they are no longer useful. It's a childish behavior or something else, I don't know

I can remember retracting my expired clothes from the trash or even from a burning fire, so that I will save them for remembrance in the near future. *(That's stupid right? ***Childish behavior.... hahaha...)*

Some people do a similar thing when it comes to updating their Contacts on Facebook.

Knowingly or unknowingly, their phone number on Facebook is the phone number they lost, maybe some 3years ago and no longer have access to it.

And I wonder why they are keeping it there

For example, if **Mr. A** lost his phone number and did not replace a new phone number on his profile

And somehow, the network providers' reactivated the inactive line and gave it to **Mr. B**.

If the **Mr. B** decides to use the phone number to open a Facebook account, guess what happens

Facebook will grant the access **to Mr. A's** account to the new user **Mr. B** because the password reset code for **Mr. A's** account will be sent to his old phone number which now belongs to **Mr. B**.

So, **Mr. A** may wake up one day, to discover that his name has changed to a strange name. And his profile picture has changed to a strange picture and his timeline is now empty!

At worst cases, **Mr. A** is locked out. He can no longer have access to his account.

Please, if the phone numbers in your Facebook account is a phone number that you no longer use, please, change it now before you lose your account to a novice or ordinary person.

Do not be among those people who make the mistake of entirely removing their phone numbers from Facebook.

Their reason could be because they want to hide their numbers from some set of jokers who would disturb their peace.

Since it is very vital that you have a phone number and or an email address associated with your Facebook account, do not remove your contact details, rather reset the visibility to "Only me".

By doing so, no one can get the contact details except you give it to them and you will be saving yourself some headache anytime you got locked out or when you want to reset your password or when your account got compromised.

What you can do:

=====

1. Log in to Facebook, Go to Settings > Account Settings > Personal Information. Update your contact information by adding a phone number and an email address.

   Note: Facebook will require you to enter your password before you can successfully make those changes.

SECRETS TO UN-HACKABLE FACEBOOK ACCOUNT

2. If you don't like people to see your phone number or email address, go to Settings > Privacy > Privacy Settings: scroll down, you'll find a section that says: "Your Information". Click on "Who Can See Your Contact?" Set it to "Only me".

Before adding any phone number, make sure the phone number is not already used to open an account.

For example: if you have 2 Facebook accounts, and you have 2 phone numbers. It is advisable that you dedicate 1 of the phone numbers to 1st Facebook account, and the other phone number to your 2nd Facebook account. This is to avoid some technical issues, especially when the 2 accounts bear the same name.

Just like any other web program, Facebook is a website and relies on its algorithms to make decisions. A program is not a human being and can never be. So, whatever decision that it makes will never be comparable to that of human beings.

Even if you decided to use a single phone number on two different Facebook accounts, a human being will know that the two accounts belongs to you, but Facebook algorithm are not designed to work that way.

Assuming you have phone number 001 in your first Facebook account, once you add the same phone number 001 to your second Facebook account as a primary phone number, one of the accounts will override the other.

In fact, the accounts will be interfering when you want to login or change your password.

If such thing happened to you, an alternative method to login is to use your username and password.

SECRETS TO UN-HACKABLE FACEBOOK ACCOUNT

But if the username for the two accounts are the same and they have the same phone number too, you are left with only one and a last option.

Use your Facebook email id and password to log in.

To find your email id, go to Facebook through a web browser, and click on your profile, on the "About" section, click on "More", you will see a button that asks you to copy link to profile.

When I clicked on mine, it showed me this:
https://facebook.com/goodness.kosi

That last part is my Facebook email-id.

Every account on Facebook has a unique email-id.

Now, if I want to login through my account, I will rewrite the last part of that URL to look like an email address.

Here it is:

/goodness.kosi will be rewritten as: goodness.kosi@facebook.com.
That's my Facebook email.

I can simply log in to my account by combining my password and goodness.kosi@facebook.com

So, if you are locked out due to complications. Follow the above process to get your email id, log in and correct the complications.

I want to draw your attention to a very important thing which I mentioned earlier.

Notice that the above email looks like a strong password but please, do not use your email as a password. It's not advisable.

Chapter 8
Take Advantage of the inbuilt Facebook security system

Certainly, the more technology opens the world, the more our privacy collapses.
Through the previous pages, I've pointed out enough measures which you can take so as to ensure that your account is safe, but there are still extra things you can do.

Facebook have an in-built function called **"Two-factor Authentication"**.

What it does is that whenever a new device tries to log into your account, a 6 digit code will be sent to your mobile number or email. And the person trying to log in will not succeed if he doesn't have access to your mobile number or email, because Facebook will ask him/her to provide those codes before he can proceed.

This a nice approach, but with its own defects.

Notice that I said a *new* device. This means that if at any time, you have used a device to log in to your account; the Two Factor Authentication won't work if you try to log in through that device again. Except if you cleared the cookies in the device or maybe you Logged out of all session as suggested earlier.

Let's get it clearer.

Assuming you enabled Two-factor Authentication on your account.

And for the first time, you took your friend's *phone* and try to log in to your account.

You typed in your username and password, and then submitted.

SECRETS TO UN-HACKABLE FACEBOOK ACCOUNT

Instead of logging in to your account, Facebook will require you to enter a 6 digit code called one time password which they sent to the phone number or email which is active on your Facebook account.

You will be able to log in after you provide the codes before it expires.

When you are done browsing your Facebook, you logged out and go.

Next time, if you use that your friend's device to log in again, Facebook will no longer send a One-time-password except you previously cleared the cookies in your friend's phone or you logged out of all active sessions from your Facebook.

Assuming that all the processes failed and your account got compromised or you got locked out; the 2nd option on Extra Security will go a long way to help.

This option allows you to choose 3 to 5 trusted friends who can help you recover your account.

Apart from enabling Two-factor Authentication, you can set up extra security like alerts so that whenever a device logs in to your account, Facebook will send you a message.

The message is usually a login notification with a link to instantly block your access if you are not the one who initiated the login.

Chapter 9

How to enable "Two-factor Authentication"

To do this, log in to Facebook, go to **"Settings > Security > Security and Login: the section: "Two-factor Authentication", click on "Use Two-factor Authentication".**

Choose a method, either a security app or text message.

----when a new device tries to log in to your account, Facebook will request a code which they will send to your number, so as to confirm that you are the one trying to login from a different device.-----

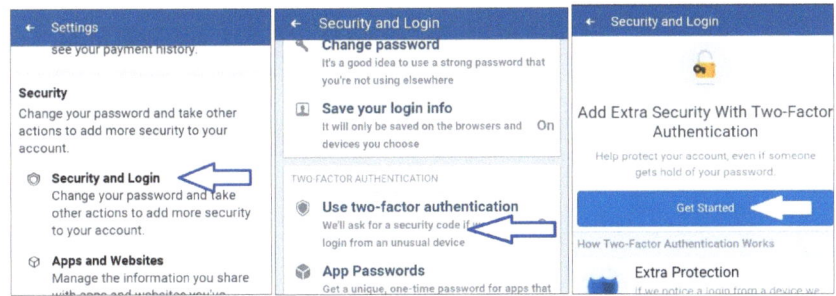

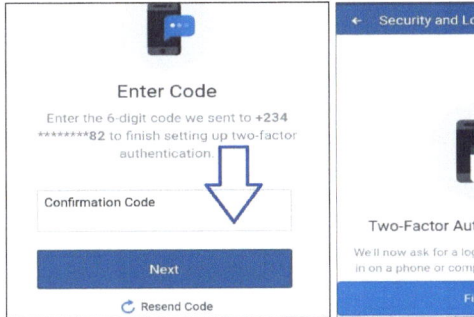

Chapter 10
How to set up extra Security

To do this, go to **"Settings > Security > Security and Login:**

Under the section: "Get alerts about unrecognized login".

Click on "Tell Facebook to notify you when someone tries to log into your account".

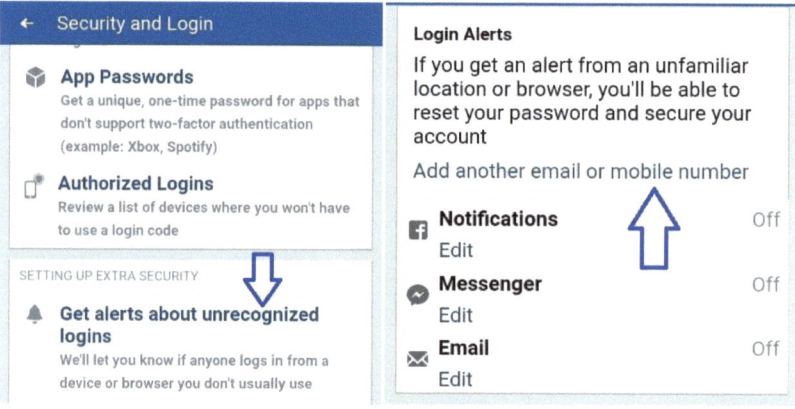

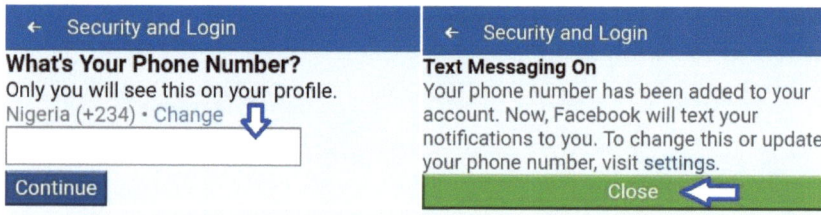

Chapter 11

How to choose 3 to 5 trusted friends who can help you recover your account

Go to **"Settings > Security > Security and Login:**

Under the section: "SETTING UP EXTRA SECURITY".

Click on "Give Facebook the names of 3 to 5 trusted friends of yours, who will be willing to help you recover your account if anything happens".

When you click on it, Facebook will ask you to select friends from your friends list.

Anytime you got locked out, you may choose this as a recovery option and Facebook will send a recovery code to one of the previously selected friends.

Note: these methods are just extra security; they are very essential and helpful. Feel free to close this book and complete these actions before continuing.

Chapter 12

Facebook Is Doing Their Part; Do Yours to Secure Your Account

There was a time that I am so jealous of Facebook that I don't want them to be as popular as they are. I was studying their technology to find out how it works. Whatever new thing I learn, I try to apply it on my own project.

But I was overwhelmed to discover that they make improvements almost every day of the week.

Facebook has continued to grow from a fun-ground to a business-environment. They know it's a very big challenge to maintain such a very big environment and they are doing everything in their power to keep it safe.

The Facebook team is ever determined to help you keep your account safe and secure.

But, you have a part to play so as to stay safe and show appreciation for their hard work.

But, is making your Facebook account un-hack-able worth the effort?

Yes, it is worth more than the effort especially if you are using your Facebook account for a serious business.

The fact is that, some people are busy having fun, others are busy making lots of money, some others are busy looking for whose advantage to take.

Nevertheless, if you follow strictly all the guidelines in this book, no hacker will be able to take over your Facebook account. Even if they do, you can easily recover it as quick as possible.

SECRETS TO UN-HACKABLE FACEBOOK ACCOUNT

Here's a summary of what you can do to make your account un-hackable Facebook account:

=====

1. Use a very strong password

2. Do not use your email, phone number or date of birth as your password.

3. A strong password should contain numerals, alphabets and symbols.

4. Do not share your password.

5. Do not click on suspicious links that asks you to login.

6. Logout and clear cookies after you log in through a device in cyber cafe or a friend's device.

7. Do not remove your phone number or email from Facebook, rather: set its privacy to "Only me".

8. Choose 3-5 trusted friends who will help you when you are locked out.

9. Enable to enable two-factor authentication.

10. And avoid the urge to hack someone's account.

11. Avoid downloading files through suspicious links

12. Avoid installing apps from non-trusted publishers.

13. Remove non-trusted or inactive 3rd party services.

SECRETS TO UN-HACKABLE FACEBOOK ACCOUNT

14. Shun password storage apps.

15. Consider changing your password at least every 6 months, although some experts recommend 3 months.

If you are able to implement 80% of these tips, I can assure you that your account will be safe and no hacker; whether ethical or non-ethical, can trespass your account, except they partner with Facebook or they broke into Facebook database, of which I know Facebook team are tirelessly protecting.

What if your account has already been hacked?

The next chapter tells you what you can do.

Chapter 13
What to Do If You're Account Was Hacked

Interestingly, it is possible to recover a hacked account! It's even easier if the phone number and email you used on your Facebook are still intact.

Even if the hacker removed the numbers and email, you can still recover the account by following the right process.

Just get your Facebook email id through the process we previously discussed. Open a browser and type in https://facebook.com/hacked.

It's even better if you do it from a device which you normally used to access the hacked account.

On the page that pops up, type in your email id and allow it to show your account.

They will ask you to type in a password even if it is your old password..

They will show your phone number or email if it is still there. Choose anyone of them to receive the code and proceed to take over your account.

If the phone number has already been replaced, tell Facebook that you no longer have access to the account.

They will require you to answer some security questions. And if you successfully got the answers without guess, they will ask you to enter your email.

Using your email, they will do the final verification and your account comes back to you if you passed the verification.

The only time you can no longer recover your account is when the hacker successfully changes your unique id.

Or when you are locked out and you fail to provide a satisfactory evidence of ownership of the account to Facebook

However, **none** of these will happen if you follow the instructions in this book.

Now, if your Facebook account is not making you any genuine money, you are missing a lot.

Evidently, there are simple processes you can follow and make a lot of money through Facebook and my new book teaches you the exact strategies

Follow me on my Facebook page so that you can get an updated once the new book is ready.

You may even get a free copy.

The End

www.ingramcontent.com/pod-product-compliance
Lightning Source LLC
Chambersburg PA
CBHW040246220526
45473CB00001B/390